HUMAN SELECTION

- Popular Science Monthly Volume 38 -

November 1890

BY

ALFRED RUSSEL WALLACE

British Library Cataloguing-in-Publication Data
A catalogue record for this book is available from the
British Library

Alfred Russel Wallace

Alfred Russel Wallace was born on 8th January 1823 in the village of Llanbadoc, in Monmouthshire, Wales.

At the age of five, Wallace's family moved to Hertford where he later enrolled at Hertford Grammar School. He was educated there until financial difficulties forced his family to withdraw him in 1836. He then boarded with his older brother John before becoming an apprentice to his eldest brother, William, a surveyor. He worked for William for six years until the business declined due to difficult economic conditions.

After a brief period of unemployment, he was hired as a master at the Collegiate School in Leicester to teach drawing, map-making, and surveying. During this time he met the entomologist Henry Bates who inspired Wallace to begin collecting insects. He and bates continued exchanging letters after Wallace left teaching to pursue his surveying career. They corresponded on prominent works of the time such as Charles Darwin's *The Voyage of the Beagle* (1839) and Robert Chamber's *Vestiges of the Natural History of Creation* (1844).

Wallace was inspired by the travelling naturalists of the day and decided to begin his exploration career collecting specimens in the Amazon rainforest. He explored the Rio Negra for four years, making notes on the peoples and languages he encountered as well as the geography, flora, and fauna. On his return voyage his ship, Helen, caught fire and he and the crew were stranded for ten days before being picked up by the Jordeson, a brig travelling from Cuba to London. All of his specimens aboard Helen had been lost.

After a brief stay in England he embarked on a journey to the Malay Archipelago (now Singapore, Malaysia, and Indonesia). During this eight year period he collected more than 126,000 specimens, several thousand of which represented new species to science. While travelling, Wallace refined his thoughts about evolution and in 1858 he outlined his theory of natural selection in an article he sent to Charles Darwin. This was published in the same year along with Darwin's own theory. Wallace eventually published an account of his travels *The Malay Archipelago* in 1869, and it became one of the most popular books of scientific exploration in the 19th century.

Upon his return to England, in 1862, Wallace became a staunch defender of Darwin's landmark work *On the Origin of Species* (1859). He wrote responses to those critical of the theory of natural selection, including 'Remarks on the Rev. S. Haughton's Paper on the Bee's Cell, And on the Origin of Species' (1863) and 'Creation by Law' (1867). The former of these was particularly pleasing to Darwin. Wallace also published important papers such as 'The Origin of Human Races and the Antiquity of Man Deduced from the Theory of 'Natural Selection" (1864) and books, including the much cited *Darwinism* (1889).

Wallace made a huge contribution to the natural sciences and he will continue to be remembered as one of the key figures in the development of evolutionary theory.

Wallace died on 7th November 1913 at the age of 90. He is buried in a small cemetery at Broadstone, Dorset, England.

HUMAN SELECTION

Popular Science Monthly Volume 38

November 1890

IN one of my latest conversations with Darwin he expressed himself very gloomily on the future of humanity, on the ground that in our modern civilization natural selection had no play, and the fittest did not survive. Those who succeed in the race for wealth are by no means the best or the most intelligent, and it is notorious that our population is more largely renewed in each generation from the lower than from the middle and upper classes. As a recent American writer well puts it, "We behold the melancholy spectacle of the renewal of the great mass of society from the lowest classes, the highest classes to a great extent either not marrying or not having children. The floating population is always the scum, and yet the stream of life is largely renewed from this source. Such a state of affairs, sufficiently dangerous in any society, is simply suicidal in the democratic civilization of our day."[1]

That the check to progress here indicated is a real one few will deny, and the problem is evidently felt to be one of vital importance, since it lias attracted the attention of some of our most thoughtful writers, and has quite recently furnished the theme for a perfect flood of articles in our best periodicals. I propose here to consider very briefly the various suggestions made by these writers ; and afterward shall endeavor to show that when the course of social evolution shall have led to a more rational organization of society, the problem will receive its final solution by the action of physiological and social agencies, and in perfect harmony with the highest interests of humanity.

Before discussing the question itself, it will be well to consider whether there are in fact any other agencies than some form of selection to be relied on. It has been generally accepted hitherto that such beneficial influences as education, hygiene, and social refinement had a cumulative action, and would of themselves lead to a steady improvement of all civilized races. This view rested on the belief that whatever improvement was effected in individuals was transmitted to their progeny, and that it would be thus possible to effect a continuous advance in physical, moral, and intellectual qualities without any selection of the better or elimination of the inferior types. But of late years grave doubts have been thrown on this view, owing chiefly to the researches of Galton and Weismann as to the fundamental causes to which heredity is due. The balance of opinion among physiologists now seems to be against the heredity of any qualities acquired by the individual after birth, in which case the question we are discussing will be much simplified, since we shall be limited to some form of selection as the only possible means of improving the race.

In order to make the difference between the two theories clear to those who may not have followed the recent discussions on the subject an illustration may be useful. Let us suppose two per- sons, each striving to produce two distinct types of horse the cart-horse and the racer from the wild prairie horses of America, and that one of them believes in the influence of food and training, the other in selection. Each has a lot of a hundred horses to begin with, as nearly as possible alike in quality. The one who trusts to selection at once divides his horses into two lots, the one stronger and heavier, the other lighter and more active, and, breeding from these, continually selects, for the parents of the succeed- ing generation, those which most nearly approach the two types required. In this way it is perfectly certain that in a comparatively short period thirty or forty years perhaps he would be able to produce two very distinct forms, the one a very fair race- horse, the other an equally good specimen of a cart-horse ; and he could do this without subjecting the two strains to any difference of food or training, since it is by selection alone that our various breeds of domestic animals have in most cases been produced.

On the other hand, the person who undertook to produce similar results by food and training alone, without allowing selection to have any part in the process, would have to act in a very different manner. He would first divide his horses into two lots as nearly as possible identical in all points, and thereafter subject the one lot to daily exercise in drawing loads at a slow pace, the other lot to equally constant exercise in running, and he might also supply them with different kinds of food if he thought it calculated to aid in producing the required effect. In each successive generation he must make no selection of the swiftest or the strongest, but must either keep the whole progeny of each lot, or carefully choose an average sample of each to be again subjected to the same discipline. It is quite certain that the very different kinds of exercise would have some effect on the individuals so trained, enlarging and strengthening a different set of muscles in each, and if this effect were transmitted to the off- spring, then there ought to be in this case also a steady advance toward the racer and the cart-horse type. Such an experiment, however, has never been tried, and we can not therefore say positively what would be the result ; but those who accept the theory of the non-heredity of acquired characters would predict with confidence that after thirty or forty generations of training with- out selection, the last two lots of colts would have made little or no advance toward the two types required, but would be practically indistinguishable.

It is exceedingly difficult to find any actual cases to illustrate this point, since either natural or artificial selection has almost always been present. The apparent effects of disuse in causing the diminution of certain organs, such as the reduced wings of some birds in oceanic islands and the very sinall or aborted eyes of some of the animals inhabiting extensive caverns, can be as well explained by the withdrawal of the cumulative agency of natural selection and by economy of growth, as by the direct effects of disuse. The following facts, however, seem to show that special skill derived from practice, when continued for several generations, is not inherited, and does not therefore tend to increase. The wonderful skill of most of the North American Indians in following a trail by indications quite imperceptible to the ordinary European has been dwelt upon by many writers, but it is now admitted that the white trappers equal and often excel them, though these trappers have in almost every case acquired their skill in a comparatively short period, without any of the inherited experience which might belong to the Indian.

Again, for many generations a considerable portion of the male population of Switzerland have practiced rifle-shooting as a national sport, yet in international contests they show no marked superiority over our riflemen, who are, in a large proportion, the sons of men who never handled a gun, Another case is afforded by the upper classes of this country, who for many generations have been educated at the universities, and have had their classical and mathematical abilities developed to the fullest extent by rivalry for honors. Yet now, that for some years these institutions have been opened to dissenters whose parents usually for many generations have had no such training, it is found that these dissenters carry off their full share or even more than their share of honors. We thus see that the theory of the non-heredity of acquired characters, whether physical or mental, is supported by a considerable number of facts, while few if any are directly opposed to it. We therefore propose to neglect the influence of education and habit as possible factors in the improvement of our race, and to confine our argument entirely to the possibility of improvement by some form of selection.*

Among the modern writers who have dealt with this question the opinions of Mr. Galton are entitled to be first considered, be- cause he has studied the whole subject of human faculty in the most thorough manner, and has perhaps thrown more light upon it than any other writer. The method of selection by which he has suggested that our race may be improved is to be brought into action by means of a system of marks for family merit, both as to health, intellect, and morals, those individuals who stand high in these respects being encouraged to marry early by state endowments sufficient to enable the young couples to make a start in life. Of all the proposals that have been made tending to the systematic improvement of our race, this is one of the least objectionable, but it is also, I fear, among the least effective. Its tendency would undoubtedly be to increase the number and to raise the standard of our highest and best men, but it would at the same time leave the bulk of the population unaffected, and would but slightly diminish the rate at which the lower types tend to supplant or to take the place of the higher. What we want is, not a higher standard of perfection in the few, but a higher average, and this can best be produced by the elimination of the lowest of all and a free intermingling of the rest.

Something of this kind is proposed by Mr. Hiram M. Stanley in his article on Our Civilization and the Marriage Problem, already referred to. This writer believes that civilizations perish because, as wealth and art increase, corruption creeps in, and the new generations fail in the work of progress because the renewal of individuals is left chiefly to the unfit.

Those who desire more information on this subject should read Wcismann's Essays on Heredity.

The two great factors which secure perfection in each

animal race sexual selection by which the fit are born, and natural selection by which the fittest survive both fail in the case of mankind, among whom are hosts of individuals which in any other class of beings would never have been born, or, if born, would never survive. He argues that, unless some effective measures are soon adopted and strictly en- forced, our case will be irremediable ; and, since natural selection fails so largely, recourse must be had to artificial selection. " The drunkard, the criminal, the diseased, the morally weak should never come into society. Not reform but prevention should be the cry." The method by which this is proposed to be done is hinted at in the following passages : " In the true golden age, which lies not behind but before us, the privilege of parent- age will be esteemed an honor for the comparatively few, and no child will be born who is not only sound in body and mind, but also above the average as to natural ability and moral force " ; and again, " The most important matter in society, the inherent quality of the members which compose it, should be regulated by trained specialists."

Of this proposal and all of the same character we may say, that nothing can possibly be more objectionable, even if we admit that they might be effectual in securing the object aimed at. But even this is more than doubtful ; and it is quite certain that any such interference with personal freedom in matters so deeply affecting individual happiness will never be adopted by the majority of any nation, or if adopted would never be submitted to by the minority without a life-and-death struggle.

Another popular writer of the greatest ability and originality, who has recently given us his solution of the problem, is Mr. Grant Allen. His suggestion is in some respects the very reverse of the last, yet it is, if possible, even more objectionable. Instead of any interference with personal freedom, he proposes the entire abolition of legal restrictions as to marriage, which is to be a free contract to last only so long as either party desires. This alone, however, would have no effect on race-improvement, except probably a prejudicial one. The essential part of his method is, that girls should be taught, both by direct education and by the influence of public opinion, that the duty of all healthy and intellectual women is to be the mothers of as many and as perfect children as possible. For this purpose they are recommended to choose as temporary husbands the finest, healthiest, and most intellectual men, thus insuring a variety of combinations of parental quali- ties which would lead to the production of offspring of the highest possible character and to the continual advancement of the race.*

I think I have fairly summarized the essence of Mr. Grant Allen's proposal, which, though enforced with all his literary skill and piquancy of illustration, can, in my opinion, only be fitly de- scribed by the term already applied to it by one of his reviewers, " detestable." It purports to be advanced in the interests of the children and of the race ; but it would necessarily impair that family life and parental affection which are the prime essentials to the well-being of children ; while, though it need not necessarily produce, it would certainly favor, the increase of pure sensualism, the most degrading and most fatal of all the qualities that tend to the deterioration of races and the downfall of nations. One of the modern American advocates of greater liberty of divorce, in the interest of marriage itself, thus admirably summarises the essential characteristics and purport of true marriage : " In a true relation, the chief object is the loving companionship of man and woman, their capacity for mutual help and happiness, and for the development of all that is noblest in each other. The second object is the building up a home and family, a place of rest, peace, security, in which child-life can bud and blossom like flowers in the sunshine." f For such rest, peace, and security, permanence is essential. This permanence need not be attained by rigid law, but by the influence of public opinion, and, more surely still, by those deep-seated feelings and emotions which, under favorable conditions, render the marriage tie stronger and its influence more beneficial the longer it endures. To me it appears that no system of the relations of men and women could be more fatal to the happiness of individuals, the well-being of children, or the advancement of the race, than that proposed by Mr. Grant Allen.

Before proceeding further with the main question it is necessary to point out that, besides the special objections to each of the proposals here noticed, there is a general and fundamental objection. They all attempt to deal at once, and by direct legislative enactment, with the most important and most vital of all human relations, regardless of the fact that our present phase of social development is not only extremely imperfect but vicious and rot- ten at the core. How can it be possible to determine and settle the relations of women to men which shall be best alike for individuals and for the race, in a society in which a very large pro- portion of women are obliged to work long hours daily for the barest subsistence, while another large proportion are forced into more or less uncongenial marriages as the only means of securing some amount of personal independence or physical well-being?

See The Girl of the Future, in The Universal Review, May, 1890, and a previous article entitled Plain Words on the Woman Question, in the Fortnightly Review, October, 1889. f Elizabeth Cady Stanton in the Arena, April, 1890.

Let any one consider, on the one hand, the lives of the wealthy as portrayed in the society newspapers and even in the advertisements of such papers as The Field and The Queen, with their endless round of pleasure and luxury, their almost inconceivable wastefulness and extravagance, indicated by the cost of female dress and such facts as the expenditure of a thousand pounds on the flowers for a single entertainment; and, on the other hand, the terrible condition of millions of workers men, women, and children as detailed in the Report of the Lords Commission on Sweating, on absolutely incontestable evidence, and the still more awful condition of those who seek work of any kind in vain, and, seeing their children slowly dying of starvation, are driven in utter helplessness and despair to murder and suicide. Can any thoughtful person admit for a moment that, in a society so constituted that these overwhelming contrasts of luxury and privation are looked upon as necessities, and are treated by the Legislature as matters with which it has practically nothing to do, there is the smallest probability that we can deal successfully with such tremendous social problems as those which involve the marriage tie and the family relation as a means of promoting the physical and moral advancement of the race ? What a mockery to still further whiten the sepulchre of modern society, in which is hidden " all manner of corruption," with schemes for the moral and physical advancement of the race !

It is my firm conviction, for reasons which I shall state presently, that when we have cleansed the Augean stable of our existing social organization, and have made such arrangements that all shall contribute their share of either physical or mental labor, and that all workers shall reap the full reward of their work, the future of the race will be insured by those laws of human development that have led to the slow but continuous advance in the higher qualities of human nature. When men and women are alike free to follow their best impulses ; when idleness and vicious or useless luxury on the one hand, oppressive labor and starvation on the other, are alike unknown ; when all receive the best and most thorough education that the state of civilization and knowledge at the time will admit ; when the standard of public opinion is set by the wisest and the best, and that standard is systematically inculcated on the young ; then we shall find that a system of selection will come spontaneously into action which will steadily tend to eliminate the lower and more degraded types of man, and thus continuously raise the average standard of the race. I there- fore strongly protest against any attempt to deal with this great question by legal enactments, or by endeavoring to modify public opinion as to the beneficial character of monogamy and permanence in marriage. That the existing popular opinion is the true one is well and briefly shown by Miss Chapman in a recent number of Lippincott's Magazine ; and as her statement of the case expresses my own views, and will, I think, be approved by most thinkers on the subject, I here give it :

1. Nature plainly indicates permanent marriage as the true human relation. The young of the human pair need parental care and supervision for a great number of years.

2. Instinct is strongly on the side of indissoluble marriage. In proportion as men leave brutedom behind and enter into the fullness of their human heritage, they will cease to tolerate the idea of two or more living partners.

3. History shows conclusively that where divorce has been easy, licentiousness, disorder, and often complete anarchy have prevailed. The history of civilization is the history of advance in monogamy, of the fidelity of one man to one woman, and one woman to one man.

4. Science tells the same tale. Physiology and hygiene point to temperance, not riot. Sociology shows how man, in spite of himself, is ever striving, through lower forms, upward, to the monogamic relation.

5. Experience demonstrates to every one of us, individually, the superiority of the indissoluble marriage. "We know that, speaking broadly, marriages turn out well or ill in proportion as husband and wife are let me not say loving but loyal, sinking differences and even grievances for the sake of children and for the sake of example.

We have now to consider what would be the probable effect of a condition of social advancement, the essential characteristics of which have been already hinted at, on the two great problems the increase of population, and the continuous improvement of the race by some form of selection which we have reason to believe is the only method available. In order to make this clear, however, and in order that we may fully realize the forces that would come into play in a just and rational state of society, such as may certainly be realized in the not distant future, it will be necessary to have a clear conception of its main characteristics. For this purpose, and without committing myself in any way to an approval of all the details of his scheme, I shall make use of Mr. Bellamy's clear and forcible picture of the society of the future, as he supposes it may exist in America in little more than a century hence.*

The essential principle on which society is supposed to be founded is that of a great family. As in a well-regulated modern family, the elders, those who have experience of the labors, the duties, and the responsibilities of life, determine the general mode of living and working, with the fullest consideration for the convenience and real well-being of the younger members, and with a recognition of their essential independence.

Looking Backward. See especially chapters vii, ix, xii, and xxv.

As in a family, the same comforts and enjoyments are secured to all, and the very idea of making any difference in this respect to those who from mental or physical disability are unable to do so much as others, never occurs to any one, since it is opposed to the essential principles on which a true society is held to rest. As regards education all have the same advantages, and all receive the fullest and best training, both intellectual and physical ; every one is encouraged to follow out those studies or pursuits for which they are best fitted, or for which they exhibit the strongest inclination. This education, the complete and thorough training for a life of usefulness and enjoyment, continues in both sexes till the age of twenty-one (or thereabouts), when all alike, men and women, take their place in the ranks of the industrial army in which they serve for three years. During the latter years of their education, and during the succeeding three years of industrial service, every opportunity is given them to see and understand every kind of work that is carried on by the community, so that at the end of the term of probation they can choose what department of the public service they prefer to enter. As every one men, women, and children alike receive the same amount of public credit their equal share of the products of the labor of the community, the attractiveness of various pursuits is equalized by differences in the hours of labor, in holidays, or in special privileges attached to the more disagreeable kinds of necessary work, and these are so modified from time to time that the volunteers for every occupation are always about equal to its requirements. The only other essential feature that it is necessary to notice for our present purpose is the system of grades, by which good conduct, industry, and intelligence in every department of industry and occupation are fully recognized, and lead to appointments as overseers, superintendents, or general managers, and ultimately to the highest offices of the state. Every one of these grades and appointments is made public ; and as they constitute the only honors and the only differences of rank, with corresponding insignia and privileges, in an otherwise equal body of citizens, they are highly esteemed, and serve as ample inducements to industry and zeal in the public service.

At first sight it may appear that in any state of society whose essential features were at all like those here briefly outlined, all the usual restraints to early marriage as they now exist would be removed, and that a rate of increase of the population unexampled in any previous era would be the result, leading in a few generations to a difficulty in obtaining subsistence, which Mai thus has shown to be the inevitable result of the normal rate of increase of mankind when all the positive as well as the preventive checks are removed. As the positive checks which may be briefly summarized as war, pestilence, and famine are supposed to be non-existent, what, it may be asked, are the preventive checks which are suggested as being capable of reducing the rate of increase within manageable limits ? This very reasonable question I will now endeavor to answer.

The first and most important of the checks upon a too rapid increase of population will be the comparatively late average period of marriage, which will be the natural result of the very conditions of society, and will besides be inculcated during the period of education, and still further enforced by public opinion. As the period of systematic education is supposed to extend to the age of twenty-one, up to which time both the mental and physical powers will be trained and exercised to their fullest capacity, the idea of marriage during this period will rarely be entertained. During the last year of education, however, the subject of marriage will be dwelt upon, in its bearing on individual happiness and on social well-being, in relation to the welfare of the next generation and to the continuous development of the race. The most careful and deliberate choice of partners for life will be inculcated as the highest social duty; while the young women will be so trained as to look with scorn and loathing on all men who in any way willfully fail in their duty to society on idlers and malingerers, on drunkards and liars, on the selfish, the cruel, or the vicious. They will be taught that the happiness of their whole lives will depend on the care and deliberation with which they choose their husbands, and they will be urged to accept no suitor till he has proved himself to be worthy of respect by the place he holds and the character he bears among his fellow- laborers in the public service.

Under social conditions which render every woman absolutely independent, so far as the necessaries and comforts of existence are concerned, surrounded by the charms of family life and the pleasures of society, which will be far greater than anything we now realize when all possess the refinements derived from the best possible education, and all are relieved from sordid cares and the struggle for mere existence, is it not in the highest degree probable that marriage will rarely take place till the woman has had three or four years' experience of the world after leaving college that is, till the age of twenty-five, while it will very frequently be delayed till thirty or upward ? Now Mr. Galton has shown, from the best statistics available, that if we compare women married at twenty with those married at twenty-nine, the proportionate fertility is about as eight to five. But this difference, large as it is, only represents a portion of the effect on the rate of increase of population caused by a delay in the average period of marriage. For when the age of marriage is delayed the time between successive generations is correspondingly lengthened ; while a still further effect is produced by the fact that the greater the average age of marriage the fewer generations are alive at the same time, and it is the combined effect of these three factors that determines the actual rate of increase of the population.*

But there is yet another factor tending to check the increase of population that would come into play in a society such as we have been considering. In a remarkable essay on the Theory of Population, Herbert Spencer has shown, by an elaborate discussion of the phenomena presented by the whole animal kingdom, that the maintenance of the individual and the propagation of the race vary inversely, those species and groups which have the shortest and most uncertain lives producing the greatest number of offspring ; in other words, individuation and reproduction are antagonistic. But individuation depends almost entirely on the development and specialization of the nervous system, through which, not only are the several activities and co-ordinations of the various organs carried on, but all advance in instinct, emotion, and intellect is rendered possible. The actual rate of increase in man has been determined by the necessities of the savage state, in which, as in most animal species, it has usually been only just sufficient to maintain a limited average population. But with civilization the average duration of life increases, and the possible increase of population under favorable conditions becomes very great, because fertility is greater than is needed under the new conditions. The advance in civilization as regards the preservation of life has in recent times become so rapid, and the increased development of the nervous system has been limited to so small a portion of the whole population, that no general diminution in fertility has yet occurred. That the facts do, however, accord with the theory is indicated by the common observation that highly intellectual parents do not as a rule have large families, while the most rapid increase occurs in those classes which are engaged in the simpler kinds of manual labor. But in a state of society in which all have their higher faculties fully cultivated and fully exercised throughout life, a slight general diminution of fertility would at once arise, and this diminution, added to that caused by the later average period of marriage, would at once bring the rate of increase of population within manageable limits. The same general principle enables us to look forward to that distant future when the world will be fully peopled, in perfect confidence that an equilibrium between the birth and death rates will then be brought about by a combination of physical and social agencies, and the bugbear of over-population become finally extinct.

There now only remains for consideration the means by which, in such a society, a continuous improvement of the race could be brought about, on the assumption that for this purpose education is powerless as a direct agency, since its effects are not hereditary, and that some form of selection is an absolute necessity. This improvement I believe will certainly be effected through the agency of female choice in marriage. Let us, therefore, consider how this would probably act.

It will be generally admitted that, although many women now remain unmarried from necessity rather than from choice, there are always a considerable number who feel no strong inclination to marriage, and who accept husbands to secure a subsistence or a home of their own rather than from personal affection or sexual emotion. In a society in which women were all pecuniarily in- dependent, were all fully occupied with public duties and intellectual or social enjoyments, and had nothing to gain by marriage as regards material well-being, we may be sure that the number of the unmarried from choice would largely increase. It would probably come to be considered a degradation for any woman to marry a man she could not both love and esteem, and this feeling would supply ample reasons for either abstaining from marriage altogether or delaying it till a worthy and sympathetic husband was encountered. In man, on the other hand, the passion of love is more general, and usually stronger ; and as in such a society as is here postulated there would be no way of gratifying this passion but by marriage, almost every woman would receive offers, and thus a powerful selective agency would rest with the female sex. Under the system of education and of public opinion here suggested there can be no doubt how this selection would be exercised. The idle and the selfish would be almost universally rejected. The diseased or the weak in intellect would also usually remain unmarried ; while those who exhibited any tendency to insanity or to hereditary disease, or who possessed any congenital deformity, would in hardly any case find partners, because it would be considered an offense against society to be the means of perpetuating such diseases or imperfections.

We must also take into account a special factor hitherto, I believe, unnoticed in this connection, that would in all probability intensify the selection thus exercised. It is well known that females are largely in excess of males in our existing population, and this fact, if it were a necessary and permanent one, would tend to weaken the selective agency of women, as it undoubtedly does now.

A Theory of Population deduced from the General Law of Animal Fertility. Republished from the Westminster Review for April, 1852.

But there is good reason to believe that it will not be a permanent feature of our population. The births always give a larger proportion of males than females, varying from three and a half to four per cent. But boys die so much more rapidly than girls that when we include all under the age of five the numbers are nearly equal. For the next five years the mortality is nearly the same in both sexes ; then that of females preponderates up to thirty years of age ; then up to sixty that of men is the larger, while for the rest of life female mortality is again greatest. The general result is that at the ages of most frequent marriage from twenty to thirty-five females are between eight and nine per cent in excess of males. But during the ages from five to thirty-five we find a wonderful excess of male deaths from two preventible causes

" accident " and " violence." For the year 1888 the deaths from these causes in England and Wales were as follows :

Males (0 to 35 years), 4,158. Females (5 to 35 years), 1,100.*

Here we have an excess of male over female deaths in one year of 3,058, all between the ages of five and thirty-five, a very large portion of which is no doubt due to the greater risks run by men and boys in various industrial occupations. In a state of society in which the bulk of the population were engaged in industrial work it is quite certain that almost all these deaths would be pre- vented, and thus bring the male population more nearly to an equality with the female. But there are also many unhealthy employments in which men are exclusively engaged, such as the grinders of Sheffield, the white-lead manufacturers, and many others ; and many more men have their lives shortened by labor in unventilated workshops, to say nothing of the loss of life in war. When the lives of all its citizens are accounted of equal value to the community, no one will be allowed to suffer from such preventible causes as these ; and this will still further reduce the mortality of men as compared with that of women. On the whole, then, it seems highly probable that in the society of the future the superior numbers of males at birth will be maintained throughout life, or, at all events, during what may be termed the marriageable period. This will greatly increase the influence of women in the improvement of the race. Being a minority, they will be more sought after, and will have a real choice in marriage, which is rarely the case now. This actual minority being furtlier increased by those who, from the various causes already referred to, abstain from marriage, will cause considerable numbers of men to remain permanently unmarried, and as these will consist very largely, if not almost wholly, of those who are the least perfectly developed either mentally or physically, the constant advance of the race in every good quality will be insured.

Annual Report of the Registrar General, 1888, pp. 106-7.

This method of improvement by elimination of the worst has many advantages over that of securing the early marriages of the best. In the first place, it is the direct instead of the indirect way, for it is more important and more beneficial to society to improve the average of its members by getting rid of the lowest types than by raising the highest a little higher. Exceptionally great and good men are always produced in sufficient numbers, and have always been so produced in every phase of civilization. "We do not need more of these so much as we need less of the weak and the bad. This weeding-out system has been the method of natural selection by which the animal and vegetable worlds have been improved and developed. The survival of the fittest is really the extinction of the unfit. In nature this occurs perpetually on an enormous scale, because, owing to the rapid increase of most organisms, the unfit which are yearly destroyed form a large pro- portion of those that are born. Under our hitherto imperfect civilization this wholesome process has been checked as regards mankind ; but the check has been the result of the development of the higher attributes of our nature. Humanity the essentially human emotion has caused us to save the lives of the weak and suffering, of the maimed or imperfect in mind or body. This has to some extent been antagonistic to physical and even intellectual race-improvement ; but it has improved us morally by the continuous development of the characteristic and crowning grace of our human, as distinguished from our animal, nature.

In the society of the future this defect will be remedied, not by any diminution of our humanity, but by encouraging the activity of a still higher human characteristic admiration of all that is beautiful and kindly and self-sacrificing, repugnance to all that is selfish, base, or cruel. When we allow ourselves to be guided by reason, justice, and public spirit in our dealings with our fellow-men, and determine to abolish poverty by recognizing the equal rights of all the citizens of our common land to an equal share of the wealth which all combine to produce when we have thus solved the lesser problem of a rational social organization adapted to secure the equal well-being of all, then we may safely leave the far greater and deeper problem of the improvement of the race to the cultivated minds and pure instincts of the Women of the Future. Fortnightly Review.